National Library of Australia Cataloguing-in-Publication entry Hart, Trish, author.
Are There Polar Bears Down There? / Trish Hart.
ISBN: 978-0-6480112-0-0 (paperback)

Series: Hart, Trish. Upside-Down Turn-Around ;
BIG/small
WET/dry
NIGHT/day

Published by www.trishhart.com
ABN 86411500580

www.trishhart.com

Are There Polar Bears Down There?

The Wildlife of Antarctica

by Trish Hart

Down in the Antarctic there are tall penguins,

short penguins,

albatrosses

and petrels

but are there polar bears
down there?

Down in the Antarctic
there are baby seals,

giant whales,

and fast dolphins

but there are no polar bears
down there!

So where are all the polar bears?

Up in the Arctic,
that's where!

What is my name?

I am an emperor penguin page 2

I am an Adelie penguin page 4

I am a sooty albatross page 6

I am a snow petrel page 8

We are Weddell seals page 10

I am a humpback whale page 12

We are hourglass dolphins page 14

I am a polar bear page 16

Here are two views of our planet Earth. Can you find where you live?

Polar bears live
in the Arctic,
up here.
It looks white because
it is very cold and icy.

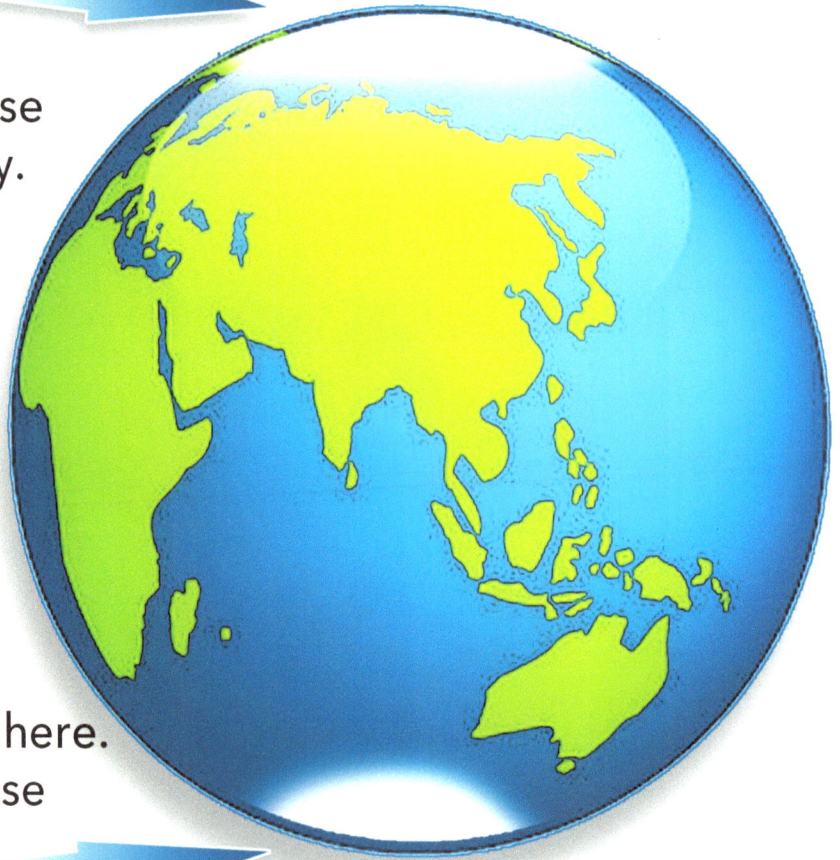

Penguins live in
the Antarctic, down here.
It looks white because
it is very cold
and icy too.

I like drawing. Do you?

Here are some of my pencil sketches of the wildlife.
Can you remember what these animals are called?

*If you like being creative, when you are old enough, you can
contact the Australian Antarctic Division and they might take you
down to the Antarctic on an Arts Fellowship, like me.
You might go on a ship like I did. It's a long way but lots of fun.
I met penguins and seals and even flew in a helicopter.
It was the best two months of my life and will always have special
memories for me.*

In memory of a little red ship, the Nella Dan.

www.ingramcontent.com/pod-product-compliance
Lightning Source LLC
Chambersburg PA
CBHW060900270326
41935CB00003B/48